ANIMALS IN THEIR HABITATS

GRASSLAND ANIMAL HABITATS

Amy Culliford

A Crabtree Roots Plus Book

Crabtree Publishing
crabtreebooks.com

School-to-Home Support for Caregivers and Teachers

This book helps children grow by letting them practice reading. Here are a few guiding questions to help the reader with building his or her comprehension skills. Possible answers appear here in red.

Before Reading:

- What do I think this book is about?
    - *I think this book is about animals.*
    - *I think this book is about animals that live in grasslands.*
- What do I want to learn about this topic?
    - *I want to learn what animals live in grasslands.*
    - *I want to know if birds live in grasslands.*

During Reading:

- I wonder why...
    - *I wonder why some animals are big and others are small.*
    - *I wonder why prairie dogs make their homes underground.*
- What have I learned so far?
    - *I have learned that giraffes live in some grasslands.*
    - *I have learned meadowlarks build nests in the grass.*

After Reading:

- What details did I learn about this topic?
    - *I have learned how streams and small rivers are in grasslands.*
    - *I have learned that many animals live in grasslands.*
- Read the book again and look for the vocabulary words.
    - *I see the word **prairie dogs** on page 8 and the word **giraffes** on page 12. The other vocabulary words are found on page 23.*

There are many kinds of animals that live in **grassland habitats**.

Some animals that live in grassland habitats are big. **Elephants** are large animals that live in some grassland habitats.

Some animals that live in grassland habitats are small.

Prairie dogs are small animals that live in some grassland habitats. They make their homes under the ground.

Many kinds of plants grow in grassland habitats. Some animals use these plants for food.

Giraffes eat many kinds of plants found in grassland habitats.

Some animals use plants to make their homes. **Meadowlarks** make their homes in the grass.

Grassland habitats have streams or small rivers. Large **ponds** are made when it rains.

Many animals drink water from these ponds.

Grassland habitats are home to many animals!

Word List

Sight Words

animals
are
big
drink
eat
food
for
found
from
grass
ground
grow
have
homes

in
it
kinds
large
live
made
make
many
of
or
plants
rains
rivers
small

some
streams
that
the
their
there
these
they
to
under
use
water
when

Words to Know

elephant

giraffe

grassland habitats

meadowlark

pond

prairie dog

ANIMALS IN THEIR HABITATS

Amy Culliford

GRASSLAND ANIMAL HABITATS

Written by: Amy Culliford
Designed by: Rhea Magaro
Series Development: James Earley
Proofreader: Kathy Middleton
Educational Consultant: Marie Lemke M.Ed.

Photographs: Gay Bumgarner / Alamy Stock Photo; pages 14-15; All other images from Shutterstock

Crabtree Publishing

crabtreebooks.com 800.387.7650

Copyright © 2025 Crabtree Publishing

All rights reserved. No part of this publication may be reproduced, stored in a retrieval system or be transmitted in any form or by any means, electronic, mechanical, photocopying, recording, or otherwise, without the prior written permission of Crabtree Publishing.

In Canada: We acknowledge the financial support of the Government of Canada through the Canada Book Fund for our publishing activities.

Printed in the USA/062024/CG20240201

Published in Canada
Crabtree Publishing
616 Welland Ave
St. Catharines, Ontario
L2M 5V6

Published in the United States
Crabtree Publishing
347 Fifth Ave
Suite 1402-145
New York, NY 10016

Library and Archives Canada Cataloguing in Publication
Available at Library and Archives Canada

Library of Congress Cataloging-in-Publication Data
Available at the Library of Congress

Hardcover: 978-1-0398-4461-2
Paperback: 978-1-0398-4542-8
Ebook (pdf): 978-1-0398-4616-6
Epub: 978-1-0398-4686-9
Read-Along: 978-1-0398-4756-9
Audio: 978-1-0398-4826-9